Anthology in the Scientific Age

Miguel A. Sanchez-Rey

Physics in the Grand Unification Scheme

Miguel A. Sanchez-Rey

Abstract

Conceptual definition of GRS and physics.

March 16th, 2016

Definition :
  GRS is *a* collection of *a* finite amount of D − variants [of stringy] in metamorphic space that can be related by their charge monopoles.

Relating one D-variant to another using their charge monopoles yields added structure [1 , 2]. In that sense not all added structures are valid. Such non-valid structures do not share opposite charge monopoles; irrelevant metamorphic energy states; and a breakdown in computational control [2, 3]. This uncontrollability yields invalid structures. Implying metamorphic space controls the D-variants by successfully merging their charge monopoles. Allowing variants to tangle and twist; to create all sorts of substructures in D-metamorphic space [2].

References

[1] Sanchez-Rey, Miguel A. The Bonding of Variants Along Their Monopoles in the Grand Unification Scheme. Vixra.org: 2015.

[2] Sanchez-Rey. Miguel A. D-variants and D-branes. Vixra.org: 2015.

[3] Sanchez-Rey, Miguel A. A Procedure in Metamorphic Space. Vixra.org: 2016.

Meta-space Energy in the Grand Unification Scheme

Miguel A. Sanchez-Rey

Abstract

What is required and restricted by the grand unification scheme when it comes to energy for any variant?

March 25th, 2016

D-variants and Meta-energy

Energy, as D-variant, has meta-space energy states such that $E = \blacktriangle + \blacktriangledown + ... \implies E = [\;\;]$. [1] Using the geometric definition of GRS [grand unification scheme] then there are a finite amount of variants [of stringy] in meta-space that are related base on their charge monopoles [2]. If variants do not follow C2R then the variants breakdown as GRS is obligated to parametrization by the TrH Theorem [3].

If $W = [\;\;] + [\;\;] + ... [\;\;]$ and $E = [\;\;] \implies W = E$ such that W is energy of meta-space state in D_v. If $p[n] \xrightarrow{\text{C2R}} M$ where M is the measure, where $M \epsilon W_{1,1}$ then $M = [\;\;] = \text{Energy} = W_{1,1}$ in terms of C2R [4].

GRS requires, in terms of energy, that meta-energy states be reducible in the Wilson operator such that the potential and kinetic energy, for superpositional quantum states, is the limit of reducibility. In that manner C2R is obeyed as long as $M = [\;\;]$ implying Energy $= W_{1,1}$.

References

[1] Sanchez-Rey, Miguel A. D-variants and D-branes. Vixra.org: 2015.

[2] Sanchez-Rey, Miguel A. Physics in the Grand Unification Scheme. Vixra.org: 2016.

[3] Sanchez-Rey, Miguel A. TrHT in the Grand Unification Scheme. Vixra.org: 2015.

[4] Sanchez-Rey, Miguel A. Current Mathematical Theory in PHPR. Vixra.org: 2015.

Infrared Divergences in the Grand Unification Scheme

Miguel A. Sanchez-Rey

Abstract

Infrared divergences are prominent in the process toward grand unification. Where do we see infrared divergences in the grand unification scheme?

April 10th, 2016

Renormalization

By the geometric definition of GRS infrared divergences can be seen in the grand unification scheme when charge monopoles are asymmetrical between D-variants [of stringy] in meta-space and at low-energy scales [1].

References

[1] Sanchez-Rey, Miguel A. Physics in the Grand Unification Scheme. Vixra.org: 2016.

PHPR in the Scientific Age

By

Miguel A. Sanchez-Rey

The Physicalist Program [PHPR] was founded when a crisis emerged at CERN in Switzerland [1]. At that moment experimental high-energy physicist sought to find SUSY by testing a variety of string models. No model had shown any consistency with theoretical SUSY as they went further in the higher-energy scale.   That said some high-energy physicist suspected that SUSY be deemed a dead enterprise. Such a possibility risked to rip the fabric of the theoretical sciences; that, is superstring theory and M-theory.   To disqualify SUSY could even mean the end of any hope of solving or testing the hierarchy problem.

Significant modifications to SUSY were sought but changing the parameters around achieved nothing and only made the experimental process much worse [1]. Some even felt that a paradigm shift in physics is imminent that could be more in line with the multi-verse of quantum physics.

Since the grand unified theory of everything had been, for many decades, the sought out attempt to merge all the natural sciences into a beautiful equation, many theories purported to solve Albert Einstein's dream; whether crude, eloquent, or elegant, these solutions all had something to contribute to furthering our knowledge about the quantum and the cosmos.   But none seem really to resolve how to unify the five forces of nature or even, and more precisely, merge quantum mechanics and general relativity into quantum gravity. The most dominant of those theories is known today as string theory.

Strings are one-dimensional objects that move around hyperspace. These objects, like knots, can be tangled, twirled, and twisted with other strings to create matter particles [fermions].   Their vibrations signify force particles [bosons]. This ultimately became 11-dimensional super-gravity.

Soon the birth of M-theory signified that five string theories be selected and that one of them be chosen to be the most likely candidate for grand unification.   Even then that theory, being the candidate, is the instantiation of all other four string theories [mirror symmetry].

With the discovery of the Higgs particle SUSY was deemed as an imminent discovery. But yet further test revealed that the Higgs is more consistent with the Standard Model than with SUSY.   That showed that there is an underlying problem with our notion of higher-energy scales at the Large Hadron Collider at CERN.

What is realized is the discovery of a variant. This variant [of stringy] is a crude variant and yet highly unique. A variant [of stringy] that yielded more variants [of stringy]. Renormalizing the variant [of stringy] led to the grand unification scheme.   This scheme consist of a set of variants [of stringy] in meta-space that can be related through their charge monopoles.

At the other end this variant [of stringy] contributed to the success of the International Thermonuclear Experimental Reactor [ITER] where a low-scale nuclear hydrogen fusion reaction was achieved. Even then ITER had limitations since even as it held promise of an endless source of energy

1

the laws of thermodynamics prohibited ITER's exploitation.   Soon a new task had to be established to resolve this new crisis. This became PHPR's First Task [1].

PHPR's First Task is a 100 year technological task.   60 percent of PHPRs 100 year task has a 40 year time-frame to be completed and another 60 years to be finally completed [2].

PHPR is design as a resolution to a foreseeable catastrophic scenario in the Scientific Age in the form of a task [2].

Before then resolving a crisis resorted to damage control and fail-safe mechanisms but since then the ever increasing complexity of advances in the technological sciences and ever greater internationalist scope of planetary society has rendered such strategic and tactical mechanisms too unstable. And so in the Scientific Age anticipating a foreseeable catastrophic scenario is the most promising strategic and tactical advantage that can handle a highly increasing technological and sophisticated society. The payoffs amount to greater technological, scientific, and mathematical advances in the form of an interplay which is the next step after the scientific method [1].

References

[1] Sanchez-Rey, Miguel A. The Physicalist Program. Createspace: 2015.

[2] Sanchez-Rey, Miguel A. PHPR. Vixra.org: 2016.

Suitablity or Non-Suitability of Certain Unification Schemes for the Grand Unification Scheme

Miguel A. Sanchez-Rey

Abstract

We aim to discuss some earlier unification candidates and whether such unification candidates are suitable or non-suitable for the grand unification scheme?

April 19th, 2016

C2R and Unification

There are six grand unified theories that are highly unique but fairly different from a variant [of stringy]. In that while variant [of stringy]'s are inherently supersymmetric they are also geometrically and algebraically one-dimensional objects that interact via their monopoles in meta-space [1]. We can also state, alternatively, that they interact via D-variant manifolds [2]. These six grand unified theories are unification schemes. Even though they comply with C2R [or the EOC Guideline] they are not essential to completing the grand unification scheme. Instead these unification schemes confirm C2R as experimentally theoretical examples [3].

These five unification schemes share similar group properties four of which can be extracted from SO [10]. Those four can be said to be minimal left-right models, SU [5], flipped SU [5], and the Pati-Salem model. The other unification scheme is the Georgi-Glashow model.

The branching rules of SO [10] are consistent with C2R. We state those rules as [1] GUT's Higg's Field [2] matter fields [3] 10 Electroweak fields. Following the direct product of those rules yields symmetry between C2R and their product values.

The Georgi-Glashow model are also consistent with C2R with the sum of the product that yields SU[5] that is keeping in mind the unitary group U[1].

Since the minimal left-right model is essentially the equivalence of left-handed and right-handed physical laws C2R is complied with base on the TrH Theorem.

Using these examples we've shown that C2R is valid. But since they are inconsistent with variant [of stringy]; in that, by variant [of stringy] we mean, and restate, one-dimensional objects that interact in metamorphic space, these unifications schemes are highly limited and inefficient. But appropriate in showing the necessity and validation of C2R.

References

[1] Sanchez-Rey, Miguel A. A Procedure in Metamorphic Space. Vixra.org: 2016.

[2] Sanchez-Rey, Miguel A. The Bonding of Variants Along Their Monopoles in the Grand Unification Scheme. Vixra.org: 2015.

[3] Sanchez-Rey, Miguel A. The EOC Guideline in PHPR. Vixra.org: 2015.

Induce Bonding Between D-variants

Miguel A. Sanchez-Rey

Abstract

What to do when there is a lack of bonding between D-variants?

May 2nd, 2016

Bonding and D-energy

D-variants bond according to the symmetrical relationship of their charge monopoles [1]. If there is very little or no bonding between them there is very little or no symmetrical relationship between their charge monopoles.

To induce bonding rather than increase D-energy change around the D-energy meta-states such that D-energy remains the same: $E = \blacktriangledown + \blacktriangle + ... \implies E = \blacktriangle + \blacktriangledown + ...$ When that happens the D-energy meta-states change such that the charge monopoles flip causing the variants to bond [2].

References

[1] Sanchez-Rey, Miguel A. The Bonding of Variants Along Their Monopoles in the Grand Unification Scheme. Vixra.org: 2015.

[2] Sanchez-Rey, Miguel A. Meta-space Energy in the Grand Unification Scheme. Vixra.org: 2016.

Detailed Analysis of Possible Adverse Consequences of the First Task of PHPR

By

Miguel A. Sanchez-Rey

Table of Contents

Introduction

The completion or non-completion of the First Task of PHPR may result in adverse consequences that could lead to the self-destruction of planetary society or an arms race between nation-states, private industries, or other planetary systems? The following concerns will be address.

Mineral Depletion and ITER

The number one goal of the First Task of PHPR is to mitigate mineral depletion as ITER [International Thermonuclear Experimental Reactor] begins to be mass-produce [1]. Mass-production begins approximately 40 years after ITER goes online [2]. Before then PHPR must be fully implemented, with a window of opportunity to complete the First Task when ITER is on the manufacturing line, and another 60 years to be fully completed.

The First Task of PHPR and the Domino Effect

After those 40 years have transpired global economic growth and environmental recovery begins to slow-down as secular stagnation dictates that continual population growth will necessitate continual stimulation of the economy and the near universal policy of recession-like economic policies. The consequences become even more dire as the ecology, at a planetary scale, begins to deteriorate due to the depletion of mineral resources and increasing population growth that results in the need for the construction of more habitats and the expansion of cities across the globe.   Though by then climate change is no longer an immediate threat the need for habitats and the expansion of cities could unleash conflict between nation-states and primitive societies in the form of territorial disputes.

These castigating catastrophic consequences, because of the laws of thermodynamics, is a domino effect that could inevitably result in a rotting and war-torn planet.

Arms and Interplanetary Conflict

Even then, with the completion of the First Task of PHPR, there is plausibility that an arms race may ensue and that the planetary state may become a threat to an advance extraterrestrial intelligence--though by theoretical estimates 99 percent of all alien intelligence has not achieved technological and scientific advances or have already destroyed themselves due to near-similar historical experiences.

There is a positive chance that an advance alien intelligence will not interfere in the affairs of Earth if and only if we reside in our solar system ands not encroach on other planetary systems. Though there is a scenario that a planetary scientific state may, concurrently with an advance alien intelligence, threaten each other with...

To avoid a global arms conflict it's imperative that the 40-year window be use productively and that means that, giving such window of opportunity, hostilities must cease between the major powers and integration of markets in the form of free-trade, and treaties that promote internationalist policies, help to federalized and pacify nation-states that cooperation becomes easier and that the Scientific Age be fully tempered as to keep the planetary system serene and quiet with ITER.

Economic Sustainability, Climate Change, and Clean-Energy

ITER, being the major catalyst, that will lead to technological achievement, by completing the First Task, will also take advantage of economic sustainability and other types of technology that inhibits climate change, and etc. But it's with ITER that these avenues are effectively pursued as ITER offers an endless source of clean-energy that will prove instrumental in environmental recovery.

Private-Markets and the Scientific Age

Private markers are to be shut-out of directly participating in the First Task of PHPR. Industries are solely motivated by short-term profits and for that reason to include the direct participation of the corporate system will result in espionage and an arms race between conflicting national interests and private interests [3]. But markets will contribute by providing the needed resources to achieve the First Task of PHPR. Eventually the disclosure of the advances made in PHPR, and the Scientific Age, will lead to acceleration in technological growth and innovation that will not deprive the general public of the wonders of scientific achievement.

References

[1] Sanchez-Rey, Miguel A. The Physicalist Program. Createspace: 2015.

[2] Sanchez-Rey, Miguel A. PHPR. Vixra.org: 2016.

[3] Sanchez-Rey, Miguel A. PHPR in the Scientific Age. Vixra.org: 2016.

Short-Term and Long-Term Strategic Concerns of the First Technological Task of PHPR

By

Miguel A. Sanchez-Rey

It's with computational control that the First Task of PHPR poses little or no risk of a large-scale runaway transformative reaction [1]. That the First Task of PHPR can be use for peaceful purposes rather than for destructive purposes. The means in which antagonist groups can acquire any technological knowledge of the First Task of PHPR is too far ahead but the case in which secrecy, for national and global security, is undeniable.

Whether the first task, which cannot be construed as a means to achieve a weapon of mass destruction, can be use to attack or harm at a low-destructive scale is a questionable matter since no known case of a habitable planet destroying itself has been observe but even then that does not mean that such scenarios are non-existent.

The first task is meant to mitigate mineral depletion but the pay-offs are limitless [2]. Even then concurrently achieving the First Task of PHPR, with ITER, will lead, at a global scale, nation-states to end hostilities and to cooperate in order to take advantage of the technological benefits of the first task – both carefully and effectively.

The technological aspects of the First Task of PHPR is to remain classified until the task is completed. The private sector is to be shut-out and the planet will remain quiet and serene until the interplay achieves its objectives. The cooperation of national governments and academic institutions will be needed. Over a period of a century technological advances and vibrancy will be accomplish.

After the First Task of PHPR is completed planetary society will have the know-how for space-travel in the form of star-gates and worm-holes which is a mile-stone achievement of the First Task of PHPR. But the human species will remain within their solar-system and seek not to interfere or colonize other outside planetary systems. Rather habitats in the form of space-stations and the large-scale use of the First Task of PHPR to change the eco-system of suitable planets within our solar system, for colonization, is the viable alternative.

There are many factors that weigh-in whether or not the First Task of PHPR will be seen as a threat to an extraterrestrial intelligence and whether or not a future militaristic society will take advantage of such technology to dominate and control. It is highly likely that such militaristic societies will destroy themselves before even considering using the First Task as an avenue for control but also its understood that the First Task of PHPR is a last-resort that may be use as détente that other planetary civilizations will not dare to encroach on Earth's natural habitat as long as Earth does not, as well, encroach on other systems.

These concerns are laid out as to ease any strategic anxiety the First Task of PHPR may have to long-term tranquility and survival. As long as limits are imposed and changes to governance is pursued any potential for a large-scale fall-out is avoidable. Rather the first task will be use as a means to preserve livelihood and survival both on Earth and in space.

References

[1] Sanchez-Rey, Miguel A. The Physicalist Program. Createspace: 2015.

[2] Sanchez-Rey, Miguel A. Detailed Analysis of Possible Adverse Consequences of the First

Task of PHPR. Vixra.org: 2016.

C2R and Anomalies

Miguel A. Sanchez-Rey

Abstract

Given a unification candidate which is both inefficient and non-optimal will the experimental physics have validity with all such candidates?

May 9th, 2016

C2R and Anomalies

If C2R is not obeyed then any unification candidate will breakdown since a variant [of stringy] can be said, in a different way, to be a unification scheme [1]. If that's the case then the variant [of stringy] does not comply with the TrH Theorem [2]. If the TrH Theorem is not complied with then the physical observables of such theories are invalid which means that, experimentally, there are anomalies with such candidates [2].

References

[1] Sanchez-Rey, Miguel A. The EOC Guideline in PHPR. Vixra.org: 2015.

[2] Sanchez-Rey, Miguel A. TrHT in the Grand Unification Scheme. Vixra.org: 2015.

The Sociopathic Norm: A Pattern of Behavior Consistent with an Unprecedented Cult of the

Scientific Age

[Author: Miguel A. Sanchez-Rey]

Table of Contents

Introduction

The Earlier Stages of the Socio-Norm: Academic and Humanitarian Cult

The Sociopathic Norm

A Potential Rabid Threat: Mass Suicide

Conclusion

Introduction

Wealth and power are the driving forces of the global market economy. That is the accumulation of wealth, on a global scale, leads to greater concentration and use of power. The more concentration of power the more likely that individuals can control their groups which preserves their wealth and privilege.

Cults are unique examples of groups with pathological, or what may be call, bizarre patterns of behavior that deviate from the norm or what is often said to be atypical. Many cults have arisen in the last century with atypical behavior; some deadly and others benign like cult followings [1, 2, 3].

Since the turn of the millennium the internet age has enabled millions to gain access to mass information. While millions of academics and institutions have also taking advantage of the internet age to promote and further their careers, intellectual property, and to enhance their status as world-changing figures. Humanitarians as well has use the internet to their advantage to promote their funds and to further their cause. These two make up the most powerful of the internet culture; which, influences all other powerful interests: politicians, actors, musicians, and etc., which feeds into the vibrancy of the internet age.

It is in this sense that the world-wide web becomes a nesting ground for cult followings as the internet tends to be a cult generator for many different aspects of academic, entertainment and social life. Whatever gets posted over the internet can generate a large cult following with a click of the mouse. Either good or favorable will generate a following as long as it sparks interests and whether or not that interest is counter-productive or not [3].

Cult icons are unique in that they have small-followings that deviates from the main-stream [3]. Their works, views, and practices are not consistent with the normalcy of social and academic thought and their methods for gaining wealth can be construed on the fringe of financial fraud.

Here we examine a new pattern of behavior consistent with an unprecedented cult of the Scientific Age. They thrive through fame and fortune but present mediocre if not ineffective or fraudulent academic and humanitarian work. They thrive by imposing brutal tactics to control each other and deceive their followers into the legitimacy and betterment of their financial and academic interests by furthering a deceptive self-image. They are atypical to such an extent that they can be said to be organize crime in the Scientific Age. Threats to the natural order and social life, and groups which will unleash whatever havoc is necessary to keep their followers from turning on them and the justice system from taking legal action against them.

They are said to be so utterly famous that that's how bad they are. And yet that's how hard it is to spot them.

The Earlier Stages of the Socio-Norm: Academic and Humanitarian Cult

It is with any suspicious behavior that law enforcement can anticipate criminal activity. Cults are no different especially with the socio-norm. Academic and humanitarian cults are pathological groups that thrive on promoting their works and foundations by disingenuous and dishonest means. That's they resort to tactical usages of academic dishonesty and financial fraud to further their careers and fames.

Academic cults resort to plagiarizing [each other if they have to], peer-reviewing each other, writing excessive amount of paperbacks and opinion papers, and furthering their ideological beliefs with limited genuine credibility to their cause. They achieve financial gains by promoting powerful interests that can feed into their credibility and further the success of their false ambitions.

Humanitarian cults resort to financial fraud usually by participating in racketeering, embezzlement and malpractice that harms those they claim to aid.

These criminal activities can be said, at a superficial level, to be bad practice that becomes bad decision-making. If an individual goes along with the members of either these two cults; then they are, in general, dishonest academics and financial frauds.

They are usually selected into either of these two groups at an early age; primarily college age, and must exhibit mediocre academic and financial performance. They are then required, after being selected, to further their criminal activity for financial and careerist gain. If they don't obey their group; then, they are brutalized rather than ostracized to protect their wealth and social-status.

Usually the selection process for induction into either of these cults comes in the way of deadly games that later maturates into sociopathic behavior. If they are able to successfully win the game; then, they are set to join the socio-norm, at an early age, that later leads to fame and fortune at the expense of genuineness and novelty.

But this is all indicative of fitting a pattern of description of dishonesty and disingenuousness. If one come's across a group of academics and humanitarians and there is observe to be suspicious behavior with either/both of them; then, there may be a sociopathic norm involve.

The Sociopathic Norm

What makes the sociopathic norm deadly is that they make sociopathic behavior into the norm by getting away with criminal decision-making that has wide-ranging catastrophic consequences to social and financial life – in other words they're sociopaths.

Their bad-decision making is an indication of superb-memorization skills but average intellectual and below-average moral intelligence in an increasingly complex and burden information age.

They thrive using the internet [especially social networking sites and blogs] and at first are considered to be leading cult figures. But by depending on each other; neglecting their differing ideological beliefs, they further their motives. They have no other motives than to further their fame and fortune and to neglect any maturity in their specialization in which they claim to be world-leaders in their prospective fields.

They depend on each other and resort to brutal tactics to control each other to the point of self-hatred and resentment. They exhibit the potential for genocidal violence but are individually inert and anti-social. They present themselves as academic and financials elitists but are in reality sycophants and cronies.

They can exhibit the charisma of invincibility as they see themselves of having the capability to get away with heinous crimes and that their network of powerful interests will shield them from the justice system.

A Potential Rabid Threat: Mass Suicide

The exposure of the socio-norm leads to the consequentiality of mass-suicide by their followers as their exposure signifies repercussions to the academic and financial system. A breakdown in the social and natural order cause by these groups will have castigating aftershocks.

To repair the damage cause by these groups the only avenue is to enforce the legal system by taking swift legal action against them. By enforcing the rule of law the first task is to identify the members of this cult; including their followers and victims, the second task is to study their academic and financial links, and the third task is to protect any remaining victims from further harm. The legal system is to then quickly breakup this cult by arresting the leading members of the socio-norm, exposing all of the members of this cult as academic and financial frauds. Eventually stripping all of them of their doctorates, title-ships, and financial assets.

At this point the sociopathic norm may become a rabid threat as whatever means will be used justify their actions, and that their own support base maybe used, as well, to incite opposition to the legal system and powerful antagonist interests will attempt to protect them.

They will use violence and social unrest to tear apart the legal system, at whatever the cause, in order to justify their actions and to preserve their reputations. The only resolution is to prosecute the members of this cult to the fullest extent to restore the natural order. Otherwise the legal system is to pursue further action that must be determine by future precedent and study.

Conclusion

The sociopathic norm can be said to be one of the most frightening cults in world-history since they are so benign and silly in nature [both in person and over the internet] that they claim genuine attributes of benevolence and altruism but are in actual malevolent and genocidal.

The sociopathic norm wants in the Scientific Age because they think they can have it all. But they end up causing conflict and chaos in the scientific process in an attempt to achieve world-fame making catastrophic decisions that tears apart the natural order.

One suspects that the sociopathic norm has existed since the beginning of the Renaissance but remain unidentifiable until the dawn of the Scientific Age. The cult went world-wide at the beginning of the world-wide web; and has become, since then, a global health and security threat.

The threat they pose are not only consequential but in those they lure into the cult. They are not to be deem initially complacent and pathological; but frightened, vulnerable, helpless, and shun at a young age. They are said, tentatively, at the time to have certain forms of minor learning and/or physical disabilities, to be in poverty, and/or a family or individual history of crime.

References

[1] Singer, Margaret Saler. Cults in our midst: The continuing fight against their hidden

menace.

[2] Moran, S. The Secret World of Cults: From Ancient Druids to Heaven's Gate. 1999.

[3] Sirkin, Mark I. Cult Involvement: A systems approach to assessment and treatment. 1990.